My Favorite COLOR

GREEN

A Crabtree Roots Book

AMY CULLIFORD

CRABTREE
Publishing Company
www.crabtreebooks.com

School-to-Home Support for Caregivers and Teachers

This book helps children grow by letting them practice reading. Here are a few guiding questions to help the reader with building his or her comprehension skills. Possible answers appear here in red.

Before Reading:

- What do I think this book is about?
    - *This book is about the color green.*
    - *This book is about things that are green.*

- What do I want to learn about this topic?
    - *I want to learn what plants are green.*
    - *I want to learn about shades of green.*

During Reading:

- I wonder why...
    - *I wonder why some frogs are green.*
    - *I wonder why grass is green.*

- What have I learned so far?
    - *I have learned that trees can be green.*
    - *I have learned that pears are green.*

After Reading:

- What details did I learn about this topic?
    - *I have learned that there are many shades of green.*
    - *I have learned that plants and animals can be green.*

- Read the book again and look for the vocabulary words.
    - *I see the word **frog** on page 7 and the word **tree** on page 8. The other vocabulary words are found on page 14.*

I see green.

I see green **grass**.

I see a green **tree**.

I see a green **frog**.

I see a green **pear**.

What do you see that is green?

Word List

Sight Words

a	I	what
do	is	you
green	see	

Words to Know

frog

grass

pear

tree

29 Words

I see green.

I see green **grass**.

I see a green **tree**.

I see a green **frog**.

I see a green **pear**.

What do you see
that is green?

CRABTREE
Publishing Company

Written by: Amy Culliford
Designed by: Rhea Wallace
Series Development: James Earley
Proofreader: Kathy Middleton
Educational Consultant:
Christina Lemke M.Ed.
Photographs:
Shutterstock: Fotofermer: cover; David Havel: p. 1; CobraCZ: p. 3; djgis: p. 4-5, 14; rsooll: p. 6, 14; LlorraineHudgins: p. 9, 14; Billion Photos: p. 11, 14; narikan: p. 13

My Favorite COLOR

GREEN

Library and Archives Canada Cataloguing in Publication
Title: Green / Amy Culliford.
Names: Culliford, Amy, 1992- author.
Description: Series statement: My favorite color | "A Crabtree roots book".
Identifiers: Canadiana (print) 20200384112 | Canadiana (ebook) 20200384120 | ISBN 9781427134677 (hardcover) | ISBN 9781427132581 (softcover) | ISBN 9781427132642 (HTML)
Subjects: LCSH: Green—Juvenile literature.
Classification: LCC QC495.5 .C852 2021 | DDC j535.6—dc23

Library of Congress Cataloging-in-Publication Data
Names: Culliford, Amy, 1992- author.
Title: Green / Amy Culliford.
Description: New York, NY : Crabtree Publishing Company, [2021] | Series: My favorite color ; a Crabtree roots book | Includes index.
Identifiers: LCCN 2020050177 (print) | LCCN 2020050178 (ebook) | ISBN 9781427134677 (hardcover) | ISBN 9781427132581 (paperback) | ISBN 9781427132642 (ebook)
Subjects: LCSH: Green--Juvenile literature. | Colors--Juvenile literature.
Classification: LCC QC495.5 .C853 2021 (print) | LCC QC495.5 (ebook) DDC 535.6--dc23
LC record available at https://lccn.loc.gov/2020050177
LC ebook record available at https://lccn.loc.gov/2020050178

Crabtree Publishing Company
www.crabtreebooks.com 1-800-387-7650

Printed in the U.S.A./022021/CG20201130

In Canada: We acknowledge the financial support of the Government of Canada through the Canada Book Fund for our publishing activities.

Published in the United States
Crabtree Publishing
347 Fifth Avenue, Suite 1402-145
New York, NY, 10016

Published in Canada
Crabtree Publishing
616 Welland Ave.
St. Catharines, Ontario L2M 5V6